IMAGES
of America

ROGER WILLIAMS
PARK ZOO

ON THE COVER: PONY RIDE ARENA, C. 1920. Human-animal interactions have always been an effective way to foster people's appreciation of—and empathy with—animals. Pony rides have been offered at different times throughout the history of Roger Williams Park, including within the zoo, at the Carousel Village (currently), or in the park itself, as pictured here. (Courtesy of Roger Williams Park Conservancy.)

IMAGES
of America

ROGER WILLIAMS PARK ZOO

Leigh Picard and Susan Ring

ARCADIA
PUBLISHING

Published by Arcadia Publishing
Charleston, South Carolina

Printed in the United States of America

Library of Congress Control Number: 2022934155

For all general information, please contact Arcadia Publishing:
Telephone 843-853-2070
Fax 843-853-0044
E-mail sales@arcadiapublishing.com
For customer service and orders:
Toll-Free 1-888-313-2665

Visit us on the Internet at www.arcadiapublishing.com

Dedicated to Sophie Danforth—founder and lifelong leader

CONTENTS

ACKNOWLEDGMENTS

We would first and foremost like to thank Emily Bisordi, whose exhaustive research formed the backbone of this book. Quite simply, this book would not exist without her contribution of knowledge, time, and energy. We would also like to thank Shareen Knowlton for her steady guiding hand throughout the creation of this book; Ron Patalano for coordinating the project in concert with all of the zoo's other ongoing projects; Tony Vecchio, former director of the zoo, for his insight and anecdotes; Renee Gamba, director of the Museum of Natural History and Planetarium at Roger Williams Park, and her staff for taking the time to search and scan their photograph archives; Brett Cortesi for digging through the zoo's old images and graphics and offering his recollections; Corrie Ignagni for poring through the zoo's collection of photographs; Corinne Nowell for seeking out additional photographs and helping to match names with faces; Sean Thomas for searching every square inch of our photographs for names, dates, and other clues; and Elizabeth Mauran for enabling us to carry this project over the finish line. Lastly, we would like to thank Dr. Jeremy Goodman, former director of the zoo, who encouraged the writing of this book to honor the zoo's 150th anniversary.

Unless otherwise noted, all images appear courtesy of Roger Williams Park Zoo.

INTRODUCTION

During the mid-19th century, the population of Providence tripled, industry grew, and people wanted refuge from a crowded, noisy city. However, there was not a lot of green space available where people could be surrounded by nature. This was true for many of the country's growing cities at that time. The urban parks movement arose out of this need and began with the development of Central Park in New York City.

In 1871, Betsey Williams, great-great-great-granddaughter of Providence founder Roger Williams, gifted 102 acres of land to the City of Providence. Visitors to the park can still see her family's farmhouse, known today as the Betsey Williams Cottage. The rural setting attracted many visitors as well as hundreds of species of wild birds and other animals—a prelude to what would become the Menagerie. In 1872, the Menagerie began within the park as a small collection of animals, including peacocks, anteaters, raccoons, guinea pigs, rabbits, hawks, and mice. The public enjoyed being in this natural setting and getting a closer look at wildlife. That same year, the park was designated as a zoo, making it one of the first zoos in the country.

In 1878, as the park acquired more land, landscape architect Horace Cleveland got approval from the City of Providence to develop the park. Development and construction continued over the years, honing and shaping the park into its current form. In 1890, the zoo's first major exhibit building, the Menagerie, was constructed. It housed large cats before being converted into an aviary and, eventually, a gift shop.

In the 1950s, Sophie Danforth visited the zoo with her four-year-old son. Feeling the zoo could do better for both animals and the community, she was spurred into action. With no previous zoo experience, and with inspiration from Ralph J. Hartman, she created the Rhode Island Zoological Society in 1963 and began a decades-long mission to raise the standards of the zoo experience for both animals and visitors.

What followed were years of development and improvement with occasional declines and setbacks. After being closed from 1978 to 1980 for improvements, a revitalized zoo opened to record crowds and enthusiastic visitors. Accreditation from the Association of Zoos and Aquariums, participation in the Species Survival Plan, and groundbreaking conservation work soon followed.

This book offers a look back at Roger Williams Park Zoo's first 150 years with an eye to its next 150.

One

1872–1899

THE EARLY YEARS

Roger Williams Park began its first animal collection in 1872, making it the third-oldest zoo in the United States. The collection consisted of small animals such as anteaters, peacocks, crows, and raccoons. In 1890, the City of Providence parks committee received $10,000 to construct the Menagerie building, which was designed with optimal features such as outdoor areas and maximum natural light for the animals. This enclosure paved the pathway for the zoo's ability to expand and grow.

In 1891, the zoo welcomed its first large animals when it received two lions from Central Park Zoo in New York City. That same year, the zoo's collection consisted of baboons, a bear, porcupines, monkeys, rabbits, deer, and prairie dogs. The real excitement came in September 1891 with the arrival of a 350-pound tiger named Prince. Within the next few years, the zoo added several other large animals to the collection, including leopards named Hamlet and Mary, jaguars named Doctor and Dora, and two lions named Palm and Rhoda.

In 1893, Roger the elephant went from being on loan to becoming a permanent member of the zoo. Around 30,000 people attended the ceremony in honor of this official announcement, and Roger remained at the zoo for 35 years.

In other milestones, the *Sentinel* dog statue became a permanent part of the park in 1896, which began its long history of being featured in photographs for generations of visitors.

ROGER WILLIAMS PARK, C. 1876. Betsey Williams, great-great-great-granddaughter of Roger Williams, the founder of Providence, Rhode Island, willed her family's 102-acre farm to the city in 1871 to honor her ancestor. One year later, the park added a small collection of animals, including raccoons, guinea pigs, white mice, squirrels, rabbits, hawks, peacocks, and anteaters, at which point Roger Williams Park officially became a zoo, making it the third-oldest zoo in the country. In 1873, the City of Cranston ceded additional acreage to the City of Providence for the expansion of the park. By 1883, the park was being developed in earnest through a series of major projects.

Menagerie Building, Exterior. The first major holding area for the zoo's collection was called the Menagerie, and although the building has been repurposed many times over the years, it still holds that name today. In 1890, the City of Providence parks committee began construction on the Menagerie for an approved amount of $10,000 with F.E. Field as the architect. The building had individual wrought-iron cages inside that were directly connected to corresponding eight-foot-square cages outside—a state-of-the-art feature for the time. The Menagerie was constructed of hard pine with asphalt on the floors and was designed so that the multitude of windows around the building would be the sole source of light above the cages. The building opened in 1891, and in the following year, the City of Providence awarded the zoo an additional $2,000 for more cages to accommodate its growing collection. (Courtesy of Roger Williams Park Museum of Natural History and Planetarium.)

Keeper Norman Robbins with Bears, c. 1900. Two baby bear cubs procured by E.L. Putnoy of Newport were brought to the zoo on October 4, 1891, after having been rescued from the poor conditions they suffered under their previous owner while in New Hampshire. They had been fed lumps of stale bread soaked in milk and were described as starving when they were removed from the box in which they had been travelling. Since the Menagerie already had a bear, the zoo operators hoped that the two youngsters could be introduced and possibly live in the same enclosure. Two additional black bear cubs, Jack and Jill, were brought to the zoo in 1904. One of several lions that lived in the zoo in the 1890s is visible in one of the enclosures in the background. The zoo's first two lions arrived from New York after having been captured in Africa. Several lion cubs were born in the following years but unfortunately had a high mortality rate.

ROGER WITH KEEPER NORMAN ROBBINS, C. 1900. Roger Williams Park Zoo has had a long line of elephants beginning with Roger, a four-year-old Asian elephant. Roger was brought to Providence in 1893 in the hopes that he would be purchased and donated to the zoo. There were no takers, so the children of Providence stepped up and raised $1,500 to purchase the pachyderm. On Arbor Day, baby Roger was formally presented to the park as a permanent member of the collection with an estimated 30,000 people attending the ceremony. Roger remained at the zoo for 35 years.

SENTINEL STATUE. The *Sentinel*, given to the City of Providence for display in Roger Williams Park in 1896, is one of the first bronze sculptures cast in the United States and the oldest outdoor public sculpture in Rhode Island. It was designed by local artist Thomas F. Hoppin and sculpted in 1851 by Thomas Frederick. It has been relocated throughout the park and zoo over its existence and is still very much a fixture within the park almost 150 years later, making the *Sentinel* the zoo's oldest resident.

CAROUSEL, C. 1900 (ABOVE) AND 2016 (BELOW). The carousel, built in 1897, opened to the public in 1899. The building, designed by Edward T. Banning, is 80 feet in diameter and constructed with a steel frame and corrugated iron roof. The horses were carved out of wood and originally had tails made from real horsehair. People gathered around this landmark for various activities such as canoeing, as shown in the lower left corner of the above photograph. In 2016, the zoo acquired the operating rights to the carousel, enabling the zoo to host parties as well as seasonal and weekly events like the hugely popular Food Truck Fridays. (Above, courtesy of Roger Williams Park Museum of Natural History and Planetarium.)

Two

1900–1959
Evolution of a Zoo

The early 1900s saw the birth of Einstein's general theory of relativity, the invention of the air conditioner, mass production of the automobile, and the Wright brothers' flight at Kitty Hawk. The Roaring Twenties brought the Jazz Age, the 1930s brought the Depression and World War II, and the 1940s brought millions of people the benefits of the GI Bill. As the decades moved into the more prosperous 1950s, television sets became a staple in more households, Elvis Presley and Marilyn Monroe were pop stars on the big screen, and a gallon of gas cost 25¢.

During this span of time, the zoo grew and changed along with these world events. By 1900, the zoo had a collection of 47 different species throughout the park. In 1906, the popular Menagerie building was closed, and there were discussions about making it into an aquarium or even an art museum. By the 1930s, however, the Menagerie was again a popular exhibit housing a large collection of birds.

In 1928, the zoo's iconic Roger the elephant was replaced by the first in a long line of Alices. She was a 24-year-old Indian elephant who stayed at the zoo until she passed away in 1967. In the late 1940s, a bear cub named Snuffy was born, the popular Bunny Island exhibit opened, and local college kids and young schoolchildren participated in a contest to submit ideas for the new children's zoo. The 1950s saw many new animals welcomed into the zoo, including llamas, a coatimundi, seals, pelicans, an emu, and a 48-pound snapping turtle that was caught in Warwick Pond by three kids.

MENAGERIE BUILDING, INTERIOR. In 1905, the park's superintendent, who did not share the same vision of expansion and growth as the previous superintendents, sold off a large portion of the Menagerie's collection—including some large cats, hyenas, and a coyote—to a private dealer. This was a common practice at the time. Shortly thereafter, the Menagerie closed, and it was used as a storage area for the next 20 years. In the 1920s, although plans were developed to turn the building into an aquarium and it did end up with several fish tanks, the plans did not come to fruition, and the Menagerie ended up housing mostly birds.

BIRDCAGES IN THE MENAGERIE BUILDING. Jump-started by a donation from local businessmen of $3,500, the initial flock included hyacinth macaws, flamingoes, mandarin ducks, and African siskins and was the largest indoor birdhouse in New England at the time. Designed to hold more than 150 birds, it measured 50 feet long, 14 feet wide, and over 14 feet high. It included natural elements such as trees and shrubbery around the periphery plus a 12-foot wading area for ducks and other waterfowl.

POSTCARD OF CAMEL WITH KEEPER NORMAN ROBBINS, C. 1900. This postcard features one of the zoo's earliest residents, a camel named Holy Moses. In 1905, the majority of animals in the Menagerie building were sold to dealers because it was difficult to give them the best of care. Holy Moses and a few smaller animals remained. Bactrian camels are still residents of the zoo today.

FEEDING CHIPMUNK, 1912. The zoo's staff has always been dedicated to caring for all creatures big and small. This photograph shows a tiny chipmunk receiving special care. During this period, the collection of small mammals and birds was exhibited throughout the park. The consolidation of exhibits into a centralized compound in 1965 made animal care, maintenance, and security much more efficient. (Courtesy of Roger Williams Park Museum of Natural History and Planetarium.)

Family Feeding Geese, c. 1920. In 1920, as today, wild geese were a fixture throughout the park. While the zoo does not officially have geese on exhibit now, it is common to see families of waterfowl roaming through the zoo. The zoo's Wetlands Trail is home to many migrating waterfowl and insects as well as other native Rhode Island species of birds, reptiles, fish, and amphibians. (Courtesy of Roger Williams Park Museum of Natural History and Planetarium.)

Early Exhibit Design. In the early 1900s, before television, the internet, or easy travel outside of the area, the ability to see an animal that one would not normally see in their own neighborhood—such as this Bactrian camel, possibly Holy Moses—was cause for excitement. Zoos of the time often exhibited animals in basic enclosures such as the one shown here. (Courtesy of Providence Public Library.)

MONKEY ISLAND. By 1900, the zoo had 47 different animal species scattered throughout the park. In 1927, thirty-six monkeys were transferred from other areas of the zoo to the newly opened Monkey Mound, later called Monkey Island. Located just to the southwest of the park's boathouse, the wide-open area contained swings, branches, and a shelter. (Right, courtesy of Providence Public Library; below, courtesy of Roger Williams Park Museum of Natural History and Planetarium.)

MONKEY ISLAND POSTCARD, C. 1920. Praised for its innovative exhibit design, the park was developing more naturalistic habitats such as Monkey Island, shown here in a postcard rendering. Instead of constraining the animals in a small cage or behind metal bars, this area was 100 feet long by 50 feet wide and had a large sunken yard surrounded by a moat, giving the monkeys a kind of free range while preventing them from escaping—or such was the plan. Despite the best intentions, 14 rhesus monkeys escaped the island in 1941. Each week, the monkeys who inhabited

the Monkey Island exhibit consumed about 10 bunches of bananas, half a crate of oranges or apples, 30 loaves of bread, and half a barrel of boiled potatoes. Unlike in today's exhibits, the public was permitted to feed the monkeys and tossed them snacks and treats throughout the day. In fact, zookeeper Bill Overton once boasted of how it set this zoo apart from others. (Courtesy of Roger Williams Park Museum of Natural History and Planetarium.)

WORKHORSE IN ALLEY, 1928. Many of the park's buildings are historic and have been used for many purposes. The barns pictured here originally housed the Roger Williams Park work animals, one of which is pictured with an unidentified human companion and several chickens. Today, these buildings house the zoo's operations and public relations offices while also serving as workshops for the zoo's graphics and facilities departments.

Sea Lion Enters Seal Pool, c. 1930. Park superintendent Martin Noonan is shown releasing a sea lion into the pool on Roosevelt Lake as an excited crowd looks on. The seal building is visible in the background. In 1946, a sea lion named Rhoda escaped from the Seal Pool and made her way toward the coast. She was subsequently spotted on a rock at Pawtuxet Cove but was never recovered. (Courtesy of Roger Williams Park Museum of Natural History and Planetarium.)

ELEPHANT BUILDING CONSTRUCTION, C. 1930. Because the zoo lacked a proper building to house elephants, the zoo's first elephant, Roger, was kept in a stall near the barn that housed the park's work animals and some of the zoo's collection animals. The first holding area built specifically for elephants was completed in 1930. It is pictured above under construction and below behind the small mammal exhibit. Like the Menagerie, the elephant building has a long history of serving different functions within the zoo. In 1980, once its purpose as an elephant barn was no longer required, it became Tropical America, an immersive, junglelike exhibit featuring windowed animal exhibits as well as free-flight birds. In 2016, it was converted into a support building for the operation of the new state-of-the-art Faces of the Rainforest exhibit, which opened in 2018.

SEAL POOL AND POSTCARD, 1938. Although it was referred to as the Seal Pool, this large area within the park was home only to sea lions, who had a daily diet of 40 pounds of fresh-caught fish. Located on one end of Roosevelt Lake, the Seal Pool was visible from a variety of vantage points along the adjacent roads. A short wall (which would have been just to the right of the area pictured below) ran from one side of the lake to the other, separating the Seal Pool from the rest of the lake. The Seal House was built in 1938 and still stands today, although the sea lions were moved into another area on zoo grounds in 1965. The Seal House building has since been refurbished and is now home to an exhibit highlighting the importance of green infrastructure and a healthy watershed. (Above, courtesy of Roger Williams Park Museum of Natural History and Planetarium.)

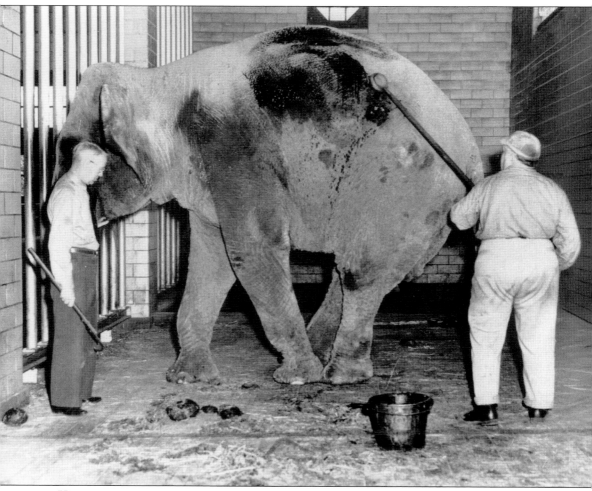

KEEPERS BATHING AN ELEPHANT, C. 1940. In 1928, Col. Joseph Samuels purchased an elephant from India and presented her as a gift to the zoo. She was described as being very gentle and tame, and there were plans to allow children to ride her. The elephant's name was Jennie upon arrival, but it was soon changed to Alice. She was the first in a long line of elephants named Alice who were kept at Roger Williams Park Zoo. (Courtesy of the *Providence Journal*.)

HARNESS RACING AT ROGER WILLIAMS PARK, C. 1930. Harness racing was a popular event at the park in the early 20th century. The track stretched through what eventually became the African area of the zoo, including the elephant exhibit and what is now the zoo's parking lot. The racetrack opened around 1910 and was used year-round. During the winter, the horses raced around the snowy track with sleighs, kicking up chunks of snow behind them. The building in the alley that housed the stables is now used as offices.

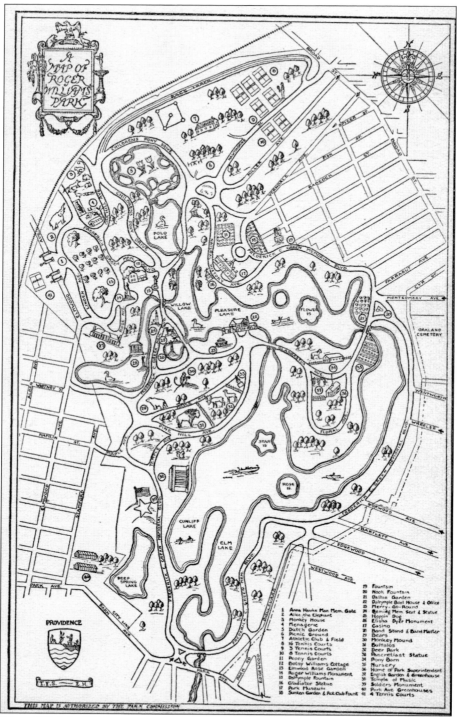

PARK MAP, 1936. In 1936, animal exhibits were located throughout Roger Williams Park. Today, the zoo occupies the northwest corner of the park from Elmwood Avenue to the tennis courts (both of which still exist today). The train tracks at the north end of the park are still in use today and run parallel to what is now Interstate 95.

28

HOPPIN

DOG AND SENTINEL, C. 1945. The statue known as *Sentinel*, designed by artist Thomas F. Hoppin, was originally displayed in front of a house Hoppin built for his new wife on the grounds of her old home, which had been destroyed by fire. Legend has it that the statue commemorates the family dog alerting the residents to the fire and that the broken chains (visible near the top of this photograph) represent his desperate efforts to rouse the family. Newspaper reports from the time, however, indicate it was the family's maid who raised the alarm, so it is unclear how much of the legend is fact.

GIRL FEEDING GEESE, 1947. While the zoo offers people of all ages an opportunity to see exotic and rare animals while learning about their history and habitats, this photograph shows that engaging with local, familiar wildlife can still bring joy and enthusiasm. The zoo's Wetlands Trail, created in 1980, sought to foster connections with nature by allowing visitors to stroll through the natural environment filled with local wetlands species. (Courtesy of Roger Williams Park Museum of Natural History and Planetarium.)

Goose Parade, 1947 (above) and 1959 (below). Although the zoo does not currently house geese as part of its animal collection, that was not always the case. The above photograph from March 1947 shows a honking parade of about 75 geese being escorted to Polo Lake. Throughout the years, the park has been a preferred location for Rhode Islanders to carry out their annual traditions, including the Goose Parade—an event during which the park's geese were led by the park superintendent from their winter holding area in the park back to the water. The tradition continued for many years, as shown in the below photograph. (Both, courtesy of Roger Williams Park Museum of Natural History and Planetarium.)

BUNNY VILLAGE, 1949. One of the zoo's most popular and endearing exhibits was Bunny Island, which opened in 1949. The exhibit featured a rabbit-scale village of human-type structures, including houses, shops, and churches. The village was appropriately decorated during the winter holidays. (Above, courtesy of Providence Public Library.)

Three

1960–1969

THE FOUNDING OF THE RHODE ISLAND ZOOLOGICAL SOCIETY

The 1960s were a milestone decade for the zoo, perhaps most significantly because Sophie Danforth founded the Rhode Island Zoological Society in 1963 and blazed the trail for the organization that runs the zoo today. She was also the zoo's president and deeply believed in its purpose and future. Danforth said that she founded the zoological society "to aid and abet the zoo, to improve it, to make it more attractive and interesting and of value educationally for the city and the state." She credited Ralph J. Hartman for originating the idea for the society, and he became its first director.

In 1963, Danforth initiated the zoo's outreach program, and for the first time, animals were brought into classrooms. She visited local schools with the zoo's screech owl and a skunk named Chanel. In 1964, two gibbons named Bob and Helen were given to the zoo by the Church Travel Agency, and Danforth picked them up in her own car. On her way back, a police officer asked her what she had in the two boxes in the back of her car, and she simply replied, "A couple of apes." Not surprisingly, the officer replied, "Oh, come on, lady," but she was prepared with certificates and soon went on her way.

ELEPHANT EXHIBIT, C. 1960. Pictures like these provide an opportunity to observe how animal care and exhibits have changed over the years. While the zoo was ahead of its time in many ways, there were still opportunities for it to evolve. For example, Roger Williams Park Zoo, like many other zoos, has moved away from exhibits that place humans at higher vantage points than animals, especially those that they would be looking up at in the wild. Likewise, safety practices have changed to the point where animal care personnel generally do not operate within the enclosure of an animal that could conceivably cause injury.

ARTISTS SKETCHING ELK, C. 1960. The Roger Williams Park Museum sponsored drawing classes and let students stroll through the park to use the natural flora and fauna as their subjects. This photograph shows two art students taking advantage of the close-up views the exhibits provided. The zoo has remained a favorite destination for artists and photographers throughout the years. (Courtesy of Roger Williams Park Museum of Natural History and Planetarium.)

MENAGERIE EXTERIOR, C. 1960. By the 1960s, the 1872 Menagerie had experienced many changes both in its collection and in its physical structure. The first noticeable change to the building's exterior was the removal of the individual cages for each of the separate exhibits. Previously, each of the apertures along the ground floor was an access point between the indoor and outdoor portions of each animal's exhibit. That outdoor space was converted, for a time, into an outdoor exhibit housing a variety of birds. As the building's function changed, the outdoor exhibits were removed in order to provide additional building access and outdoor seating.

PLAINS AREA, C. 1965. In the 1960s and 1970s, the Plains area stood in what would eventually become the Plains of Africa exhibit. At the time, there were a variety of species throughout the area, including sika deer, Barbados sheep, and eland (plains antelope). Unfortunately, not much consideration was given to viable breeding practices. In 1981, the Association of Zoos and Aquariums (AZA) founded the Species Survival Plan (SSP), a cooperative plan between institutions that serves as a powerful tool in combating extinction. Shortly after becoming the first zoo in New England to be accredited by the AZA in 1986, Roger Williams Park Zoo suspended breeding and expansion in the Plains area to develop a cohesive plan for safer and more successful breeding. Since that time, the zoo has been an active participant in the SSP for numerous species.

ROGER WILLIAMS PARK ZOO AREA, C. 1960. Before 1965, the animals were spread throughout the park; the collection was eventually consolidated as the park moved the animal exhibits within a single fenced-in compound. In this photograph, taken before the consolidation, the racing track is the most prominent landmark, located to the right of the buildings that housed the zoo's earliest animal exhibits. Also visible toward the upper left side of the image are the elephant barn, Menagerie, and stables (the long building adjacent to the road that would ultimately become Interstate 95).

SEAL POOL AT ZOO, 1965. In 1965, the zoo opened a new seal pool, this time within the confines of the zoo proper. Unlike the pool formerly located in the park, this seal pool housed both seals and sea lions. The pool was eventually converted into a penguin exhibit and then a vastly expanded and improved seal exhibit.

EVOLUTION OF EXHIBIT DESIGN. Roger Williams Park Zoo was one of the first zoos in the United States to exhibit animals in naturalistic environments and mixed-species exhibits—the above photograph of the deer park, from the late 1960s, offers an example of both. Shady trees and a wide grazing area made exhibits like this more comfortable for animals and more enjoyable for viewers. Placing different species in the same habitat, such as the deer and bison shown above, provides visual variety and a more dynamic viewing experience. In the cheetah habitat (pictured at right in 2015), animals are provided with structures for climbing and perching, giving them opportunities to get additional exercise and a bird's-eye view of their surroundings.

DOCENT VOLUNTEER, 2016. In 1966, twelve women attended a course at Brown University where they were trained to give tours of the zoo to children. The following year, they trained as the zoo's first group of docents. The zoo's Docent Council was developed in 1977. Since then, the program's training and responsibilities have evolved, and docents have contributed hundreds of thousands of volunteer hours while engaging with visitors in many capacities, including on-site interpretation, hands-on activities, event support, and close-up animal encounters. William DeNuccio, who began his docent career in 1998, is at right.

DEER EXHIBITS, 1967. Deer first became part of the zoo's collection as early as 1891 and were among the first residents of the then-new Menagerie building. Deer were also exhibited in fields along with cattle and goats in the 1920s. In 1934, park superintendent Ernest K. Thomas attempted to transfer some deer, elk, and bison to other zoos, since the cost of feeding them was becoming too expensive. Other zoos declined, citing the same challenges. In 1967, a new deer park was constructed within the zoo. Fourteen deer were tranquilized and moved to the new location.

RISPCA FLOAT WITH DIRECTORS, 1970. Roger Williams Park Zoo supported the Rhode Island Society for the Prevention of Cruelty to Animals (RISPCA) centennial celebration with this parade float. Pictured on the float are Roger Valles (left) and Dion Albach (right). Valles, who is holding a chimpanzee, was the zoo's senior keeper. In 1977, Valles became the zoo's acting director. He officially became director one year later and remained in that post until 1989. Valles was the only African American zoo director in the country at that time. Albach was the zoo's first full-time director and made many changes, including changing animal diets, creating more efficient keeper schedules, developing keeper training programs, and trading surplus animals. The Rhode Island Zoological Society provided Albach with a travel allowance, enabling him to attend conventions in Pittsburgh and Los Angeles.

Four

1970–1999
The Buddy Cianci Years

In 1986, Roger Williams Park Zoo became accredited by the Association of Zoos and Aquariums (formerly known as the American Association of Zoological Parks and Aquariums). This was a great achievement and honor, as it became the first zoo in New England to receive the accreditation, which meant that Roger Williams Park Zoo could apply for national grants and participate in programs with other AZA-accredited zoos.

Another important milestone during this time occurred in 1994, when the US Fish and Wildlife Service joined forces with the zoo to breed the American burying beetle. This hugely successful program is still going strong, with more than 5,000 beetles reared and close to 3,000 of them released in Nantucket, Massachusetts.

During the 1970s, Providence mayor Buddy Cianci knew the importance of the zoo and dedicated energy to upgrading many of the exhibits—an effort not seen since the 1930s. In 1978, the zoo was closed to accommodate large-scale construction and improvement projects, including the conversion of the elephant house into a new Tropical America exhibit, overhauling the North America exhibit area, and adding a Children's Nature Center. The zoo reopened in 1980 to great fanfare and unprecedented popularity.

PLAINS EXHIBIT WITH ZEBRA, C. 1970. Although it was hoped that zebras would be included in the zoo's collection from the time it opened in 1897, they did not arrive until 70 years later. Since this is a year-round zoo in New England, the collection must consist of robust species that can be at home in any weather. Many species that are typically thought of as warm-weather animals are actually able to adapt well to Rhode Island winters. All of the zoo's current animals have strict temperature ranges and are closely monitored by animal care staff and provided indoor access if the day's (or night's) temperature is expected to go above or below their range.

PLAINS EXHIBIT PLAY AREA, 1970s. Four decades before Hasbro's Our Big Backyard became a popular destination within the zoo, there were play opportunities for zoo visitors. The footprint and presentation of the African animal exhibits evolved dramatically over the years, but the recognition of the value of play opportunities has remained.

POLICE ANIMAL HOLDING, C. 1975. The zoo's barns were also home to the Providence Police Department's animals in addition to the park's work animals and some of the zoo's collection animals. The K-9 holding facility is pictured below with three dogs visible in the image. This facility is adjacent to the building that would eventually become the Sophie Danforth Center and house the zoo's administrative offices beginning in 1986. The longer building in the above photograph housed the police department's horses.

Keeper Joan Ferguson with Camel, c. 1975. The zoo's first female zookeeper is shown here leading a camel out of its holding area. Ferguson joined the zoo staff in 1970 as one of eight animal and bird handlers. A lifelong animal lover, she rose through the ranks to become the associate zoo director, retiring from that position in 1999. She graduated from the University of Rhode Island with a bachelor's degree in psychology. On August 26, 1970, an article in the *Providence Journal* quoted Ferguson as saying: "The trend is now toward keeping an animal in a closer approximation of its natural environment. I think someday all zoos will be designed that way. I hope so."

SEA LION PERFORMANCE, C. 1975. As part of a performance for visitors at the new seal pool located within the zoo, a sea lion was trained to announce the start of her show and dunk a basketball, among other behaviors. At the time, it was common to train animals for shows demonstrating animal behaviors as entertainment. While most of the zoo's animals are still trained today, it is no longer done as entertainment but rather to assist the animal in its own care. For example, an animal might be trained to present a body part for examination or an injection or asked to move to another location within its habitat so that keepers can safely clean the area or provide food.

ZOOMOBILE, 1978 (RIGHT) AND 2022 (BELOW). The Roger Williams Park Zoomobile has enabled the zoo to bring some members of its collection to audiences away from the zoo, including at schools, camps, and nursing homes. In more recent years, these "animal ambassadors" have consisted of a separate group of animals from those on exhibit to daily visitors. The animal ambassadors are comfortable in small groups, used to being handled by staff, and—in some cases—accustomed to being gently touched by audience members. They can include species ranging from rabbits, ferrets, and skunks to hawks, armadillos, and cockroaches.

ZOOMOBILE PROGRAMS, 1978. In addition to facilitating Zoomobile programs, Jane Demming, who is shown presenting a bird in the above photograph, also oversaw the areas of education, marketing, and public relations, among others. Eventually, each of these departments was separated and, in the case of the education department, grew to include five managers and dozens of year-round staff members. The education department grew to oversee not only Zoomobile programs but also camps, school programs, on-grounds animal encounters, exhibit and event graphics, docent volunteers, live interpretation and shows, and the Hasbro's Our Big Backyard nature play area.

POLAR BEAR EXHIBIT. The zoo closed for major renovations in 1978. It reopened in 1980 with a new Children's Nature Center, a boardwalk through a native wetlands area (which eventually became the Feinstein Junior Scholar Wetlands Trail), and a massively popular polar bear exhibit that remained a major draw until it closed in 2005. The exhibit featured a rocky landscape, waterfalls, and a pool that was visible from a viewing window below the main exhibit.

POLAR BEAR SIGN AND LOGO. The polar bear exhibit was so successful—and the polar bears were so popular—that they became the recognizable face of the zoo. The bears were so iconic that they were ultimately featured as the sole animal on the zoo's main entrance sign and remained part of the zoo's logo until almost 10 years after the exhibit was closed in 2005 to make way for a new eagle habitat as part of the zoo's master plan.

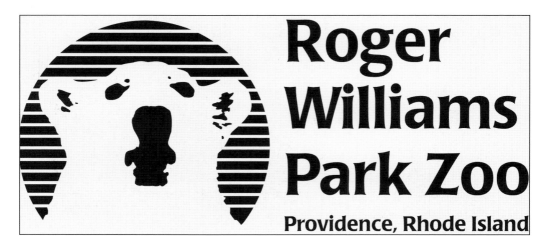

ZOO REOPENING EVENT, 1980. Sophie Danforth, founder of the Rhode Island Zoological Society, and Providence mayor Buddy Cianci spoke at the zoo's grand reopening in 1980. This era saw an unprecedented increase in improvements to the zoo. In 1986, the president of the American Association of Zoological Parks and Aquariums (later called the Association of Zoos and Aquariums) declared that Roger Williams Park Zoo was "on the verge of greatness."

ORIGINAL MAIN ENTRANCE. The zoo's original main entrance was located at the opposite end of the zoo from the current main entrance. While these gates still exist, they are no longer operational. The gates are not accessible to visitors from within the zoo but are visible from the park's Rose Garden area. While the zoo eventually created parking space for 837 cars, when these gates were the zoo's main entrance, all parking was on the street.

"CORN CRIB" CAGE, 1980s. This raccoon exhibit stood between the Menagerie and Children's Nature Center and is a prime example of how animal enclosures have evolved. Unlike in this exhibit, today's animal enclosures are designed and constructed to enhance not just the visitor's experience but—first and foremost—the animal's welfare. Round holding areas like this lack the benefits to an animal's mental and physical health that a more naturalistic habitat provides.

Feinstein Junior Scholar Wetlands Trail, 1980 (above) and 2018 (below). This quarter-mile trail, which was added to the zoo after its 1980 reopening and enhanced with above-flood-level bridges in 2007, winds through natural wetlands within the zoo's borders. There are no formal animal exhibits in this area in an effort to preserve and highlight Rhode Island's natural wildlife. Visitors can stroll along the pathway surrounded by animals and plants thriving in their natural habitat. The exhibit is filled with the sounds, sights, and splashes of its inhabitants, which range from snapping turtles to beavers to red-winged blackbirds and migrating dragonflies. Wooden easels are placed along the way so that artists can paint en plein air, and a bird blind lets visitors observe birds in a way that will not disturb them.

TROPICAL AMERICA, C. 1985. The zoo's first elephant building, constructed in 1930, was repurposed in 1980 as Tropical America and included saki monkeys, cotton-topped tamarins, and flamingos. The flamingos, housed in a separate structure from the main exhibit, were marched along the pathways to the building each morning and back to their shelter each night. The exhibit was enhanced in 2007 with the addition of new interpretive signage, improved pathways, a docent station, and a new giant anteater exhibit. The building was closed in 2016 to accommodate the new Faces of the Rainforest exhibit.

SEAL POOL EXPANSION. The harbor seal habitat was expanded in the late 1980s, which allowed for a variety of enhancements for both animal comfort and human interaction. The shape and configuration of the pool gives the animals structures to swim around and lie on, and the large deck (visible near the back of the photograph) allows for feeding opportunities and training or examination sessions.

UNDERWATER SEAL VIEWING IMPROVEMENTS. An underwater viewing area gave visitors the opportunity to observe harbor seals doing what they spend most of their time doing—swimming. When a crack was discovered in one of the viewing widows during a maintenance process in 2015, the wall was found to have other structural issues that also needed to be addressed. The pool was drained, the seals were temporarily relocated to an aquarium in New York, and work was done to replace the three smaller viewing windows with one larger one. Thanks to contributions from over 8,000 individuals in the community and an anonymous donation of $50,000, the zoo raised $125,000 for these improvements.

ROGER WILLIAMS PARK ZOO, 1985. After an enthusiastic response from the public following the zoo's transformation and reopening in 1980, the following decade continued to see significant development at the zoo. In 1983, the local chapter of the American Association of Zookeepers was founded. The zoo received accreditation from the Association of Zoos and Aquariums in 1986—a certification that still stands today. Curator Tony Vecchio directed the zoo's participation in conservation programs and continuing participation in Species Survival Plan programs. A new master plan was developed and consisted of several phases. Vecchio became the acting director of the zoo in 1989 and was installed as the official director one year later.

SOPHIE DANFORTH CENTER DEDICATION, 1986. By the mid-1960s, like so many zoos across the country, Roger Williams Park Zoo was showing visible signs of neglect. Recognizing that the zoo was an extremely valuable institution in need of organized assistance, Sophie Danforth founded the Rhode Island Zoological Society in 1962 with the purpose of increasing public awareness and support. In 1986, one of the zoo's large barns, which had been converted into administrative and support space, was renovated and rededicated as the Sophie Danforth Center. In the below photograph, Providence mayor Joseph R. Paolino Jr. is shown unveiling the dedication plaque with Danforth. In 2002, an agreement was reached with the City of Providence in which management of the zoo would transition entirely to the society.

PARK STABLE BUILDING. These two photographs show one of the zoo's early stables when it was housing work animals. Although it became the zoo's base of operations in 1986, its animal-related functions did not end there. The basement of the Sophie Danforth Center also served as the on-site veterinary facility until the zoo opened the John J. Palumbo Veterinary Hospital in 2011. Since then, that basement space has housed the zoo's animal ambassador collection. This diverse group of about 60 animals—including skunks, armadillos, hornbills, rabbits, tortoises, a red-tailed hawk, and Madagascar hissing cockroaches—is not part of the zoo's exhibit collection. Instead, the critters travel with members of the education team to off-site programs or participate in on-grounds encounters in order to provide members of the public with unique, close-up experiences.

SOPHIE DANFORTH CENTER RENOVATION. When this building was renovated in 1986, its purpose was shifted from a barn to the Sophie Danforth Center, which became the zoo's base of operations. Most of the zoo's administrative offices—including the executive director's office and the human resources, membership and development, and finance departments—are housed upstairs. The ground floor contains guest amenities such as first aid and lost and found as well as additional office and meeting space, a staff library, and the zoo's radio communication center, which keeps every staff member in constant contact with all departments of the zoo for routine day-to-day operations and in case of emergencies.

TICKET BOOTHS (ABOVE) AND STARLIGHT FESTIVAL (BELOW), 1989. When the animal exhibits were originally scattered throughout the park, admission was free. Even later, when the zoo's collection was consolidated into a more centralized location, visitors were not charged per a contract with the City of Providence. In 1986, the zoo began charging an admission fee of $2 ($1 for children under 13). As a nonprofit organization, the zoo depends on admission prices, membership fees, sponsorships, program fees, and donations to support its animals and conservation efforts. As a continued thank you to the residents whose tax dollars support the zoo, the first Saturday of every month features free admission for Providence residents.

MENAGERIE BUILDING RENOVATION. Eventually, the animal collection within the 1872 Menagerie was dispersed, and in 1989, the building was repurposed yet again. Although it was still called the Menagerie, the building was updated and turned into a café and gift shop. The building's original copper and masonry elements were restored, and the original slate roof was replaced. After the café ceased operation, the gift shop expanded, and it is still in operation today. Astute visitors will notice the access windows, located at regular intervals along the brick wall, that once connected the building's indoor and outdoor animal enclosures.

ELEPHANTS IN AFRICA EXHIBIT. The zoo's three African elephants—Ginny, Kate, and Alice—are visible at left in the above picture. Born in Gwangi National Park in Zimbabwe around 1985 and arriving at the zoo in 1990, they are unusual residents in that they were not born in a zoo. AZA-accredited zoos rarely procure animals from the wild, with rescued animals being an exception. Their herd in Zimbabwe had grown too large to be supported by the available land and was slated to be culled. Many US zoos worked to provide homes for the younger elephants—including Ginny, Kate, and Alice. After changing hands a few times, they ultimately arrived at the Roger Williams Park Zoo, where they quickly became visitor favorites.

ADAPTATIONS EXHIBIT, C. 1992. The Adaptations exhibit underwent multiple refurbishments and redesigns. After its initial phase as the Adaptations exhibit and its conversion to the Children's Nature Center in 1980, the building was reimagined as Australasia in 2000, housing wallabies, kookaburras, and snake-necked turtles. In 2016, as preparations were made to develop the land currently occupied by Tropical America into what would become the Faces of the Rainforest exhibit, Australasia was rebranded World of Adaptations and began to house animals that formerly lived in or adjacent to Tropical America. The exhibit's collection was further enhanced by the acquisition of the zoo's first Komodo dragon, making Roger Williams Park Zoo the only zoo in New England to exhibit a Komodo dragon.

TRICERATOPS, 1992. In 1992, the zoo hosted the first of its recurring dinosaur exhibits. These realistic-looking animatronic figures were placed in naturalistic habitats along the trail, awing audiences and providing a sense of how these creatures might have looked when they roamed the earth millions of years ago. The exhibit returned in 1994 and 1997.

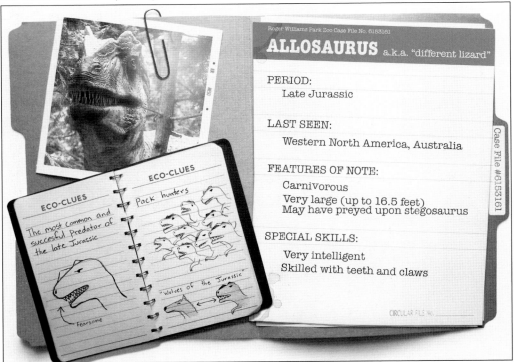

DINOSAUR IDENTIFICATION GRAPHIC, 1992. Just like it does for every other animal, the zoo designed ID graphics for the dinosaur exhibits complete with diet, range, and natural history information. In the case of the dinosaur exhibits, these signs were done in the style of a case file, featuring handwritten notes and observations.

TYRANNOSAURUS REX, 1992.
The first dinosaur exhibit served
as a fundraiser for the zoo,
with proceeds paying for the
development of the Marco Polo
Trail, which opened in 1996. As
the first exhibit to use graphics
and live interpretation to integrate
cultural storytelling with natural
history, the Marco Polo Trail
was significant for the zoo.

DINOSAURS' ARRIVAL, 1992.
Two dinosaurs, along with a
caveman passenger, are shown
arriving at the zoo for their
inclusion in the animatronic
dinosaur exhibit. As part of
the exhibit, actors portraying
roles ranging from scientists
to time-traveling explorers
engaged with visitors along the
trail, adding an extra layer of
immersion to the experience.

PLAINS OF AFRICA OPENING, 1993. The Plains of Africa opened in 1993 and underwent a major upgrade in 2006, which included the name being changed to the Fabric of Africa. The elephants' and giraffes' yards were enhanced, and the Textron Elephant and Giraffe Pavilion was renovated and expanded. Pictured above is Rodney Robinson of CLR Design, the principal architect on the project. In 2017, the zoo announced its partnership with Ivory Ella, a Rhode Island clothing company dedicated to saving elephants, and renamed the elephant yard in its honor. Designed to replicate the daily routine of wild elephants, the new elephant yard consisted of a nine-foot-deep pool for bathing, hanging feeders located near the perimeter for optimal viewing by guests, wide areas and varied topography for roaming, and a variety of enrichment items.

GIRAFFE OBSERVATION AREAS. The giraffe exhibit has always been popular, and the zoo has offered a variety of ways for visitors to observe and interact with the animals. The fencing surrounding the exhibit follows the natural terrain, providing vantage points from several elevations. Some areas provide an eye-to-eye view, while others let viewers appreciate the full height of the giraffes by placing the observer at ground level. The giraffe exhibit, similar to many other exhibits, has feeding stations along the periphery so that guests can observe the animals even more closely. Over the years, the zoo has offered different opportunities for guests, including feeding encounters where guests can provide a branch to the giraffe, getting a unique sense of the animal's size and strength, and behind-the-scenes educational programs where participants get a firsthand look at the care keepers provide.

PENGUIN EXHIBIT, 1992. The penguin exhibit, which closed in 2013, was located across from the Sophie Danforth Center in what would eventually become the location of the Alex and Ani Farmyard. A tiered seating area allowed visitors to sit comfortably for longer periods of time, encouraging active observation of the animals' behavior.

Penguin Exhibit and Identification Graphic. The penguin exhibit offered visitors experiences beyond simply viewing animals. Graphics like the one pictured below provided guests with the opportunity to become active participants as opposed to passive observers. In the case of this exhibit, signage featuring the penguins' names and their corresponding band colors encouraged visitors to try to identify the individual penguins by sight.

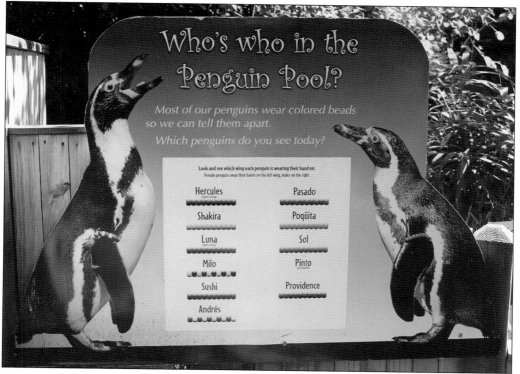

Who's who in the Penguin Pool?

Most of our penguins wear colored beads so we can tell them apart.

Which penguins do you see today?

Look and see which wing each penguin is wearing their band on:
Female penguins wear their bands on the left wing, males on the right.

Hercules (right wing)	Pasado
Shakira	Poqüita
Luna (left wing)	Sol
Milo	Pinto (no band)
Sushi	Providence
Andrés	

Trixie & Norton
(Ursus maritimus)
Proudly invite you!

Polar bear cub birth date:
November 4, 1997

Home:
Roger Williams Park Zoo
Providence, Rhode Island

POLAR BEAR BIRTH ANNOUNCEMENT, 1997. In 1997, the zoo was thrilled to announce the birth of a polar bear cub, Triton, who weighed in at less than five pounds and measured about seven inches in length. With the species considered endangered, plus a high mortality rate among newborn polar bears in zoos, the successful birth was a cause for great celebration.

Five

2000–2019
THE MODERN ZOO

These years were a time of change. The internet was well established, and Google became the go-to for just about everything. Smartphones connected people in an instant. GPS was in full force, guiding drivers along their routes, while hybrid vehicles came on the scene as environmental issues became more prominent. Things were changing at the zoo as well.

The zoo placed more focus on the visitor, with a deep dive into understanding the community's point of view in order to give people a sense of ownership and investment. The beginning of the 21st century became a time of designing exhibits that reflected what people were looking for, encouraging dialogue and audience engagement. The zoo raised the bar on what it offered, from something as simple as putting in more seating on the grounds to offering information on career development. It also became a more inclusive space, taking into consideration people with different needs and abilities. The zoo received certification from KultureCity for developing areas and activities for those with sensory accessibility needs.

This was also a time that changed the visitor experience from passive to hands-on. There were now opportunities to feed the animals, get an aerial view from a zip ride, and walk through a magnificent immersive rainforest exhibit. Hasbro's Our Big Backyard offered children an opportunity to learn through play. The zoo played a key role in environmental education with No Child Left Behind; for the first time, there was language and funding to support environmental education in public schools.

Roger Williams Park Zoo also broke ground in animal welfare, as it was the first zoo to use a sand substrate on the floor of the elephant barn. Seeing that this was highly beneficial to the animals, it paved the way for other zoos to do the same.

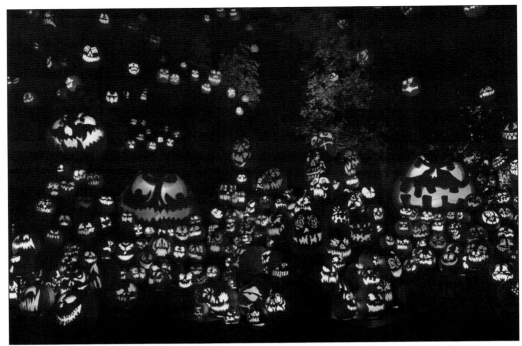

JACK-O-LANTERN SPECTACULAR. In 2002, a new tradition was born—the Jack-O-Lantern Spectacular. This annual event is so popular that it attracts visitors and news coverage from all over the world. Featuring over 6,000 individually carved and lit pumpkins, this event runs nightly throughout October. Each year's event follows a different theme—music, literature, art, and the like—and combines traditional cut-out jack-o'-lanterns with larger intricately carved pumpkins, all enhanced by music and lighting. The map shows the layout of the first Jack-O-Lantern Spectacular trail, which occupied one corner of the zoo.

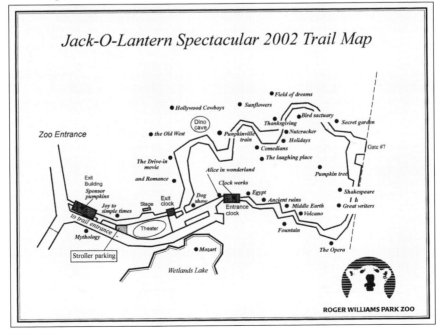

Jack-O-Lantern Spectacular Carving. Over time, the Jack-O-Lantern Spectacular grew in scope and scale, and it now encompasses the entire Wetlands Trail and surrounding paths—nearly a half mile. Passion for Pumpkins, a professional multimedia company from Oxford, Massachusetts, founded by John Reckner, produces shows around the country and designs and carves the pumpkins each year. Although the show typically displays about 6,000 pumpkins at one time, those will not last the entire show, so new pumpkins are constantly being created to replace older ones. The show goes through roughly 20,000 pumpkins during the course of its one-month run. The above photograph shows an example of one of the "intricates," while the below image shows Russ Anderson carving an image on one of the show's larger pumpkins.

Jack-O-Lantern Spectacular Work Area. Designing and planning for the show is a year-round process for Passion for Pumpkins, with installation usually beginning in August. As the zoo's busy season winds down, the Wetlands Trail is closed to the public to facilitate installation of the show's support elements, like lighting and display platforms.

Jack-O-Lantern Spectacular Pumpkin Detail. The success of the Jack-O-Lantern Spectacular, both as a fundraiser for the zoo and as a tradition for zoo visitors, has led to additional annual multiweek events such as the Holiday Light Spectacular and the Asian Lantern Spectacular. In 2019, PBS produced a documentary about the show called *Oh My Gourd! The Jack-O-Lantern Spectacular Story*. In 2021, despite the COVID-19 pandemic, the zoo set a record for attendance at the event since its debut in 2002.

ROGER WILLIAMS PARK ZOO

Itsy Bitsy Empire Logo, 2002. The press release for the 2002 opening of Itsy Bitsy Empire read: "A brand new exhibit, the Itsy Bitsy Empire, will showcase some of the world's most feared, most dangerous and most misunderstood creepy, crawly critters including tarantulas, spiders, centipedes and millipedes." Community leaders participated in an insect-cooking contest led by John Palumbo, who was then serving as chairman of the board. The stand-alone buildings that housed the one-year exhibit were originally planned to be destroyed, but one remains in operation today as the Snake Den.

Autumn Animal Enrichment. Animal care at the zoo has long included providing animals with more than simply food and shelter. Physical and mental well-being are critical in proper animal care. These can be served by providing animals with enrichment, items that provide mental stimulation, and a change in routine. This can include anything from placing new objects or smells in the enclosure to serving treats in puzzle feeders that encourage animals to use their natural foraging behaviors. In the examples pictured here, a harbor seal (above) and red panda (below) investigate pumpkins brought to the zoo as part of the Jack-O-Lantern Spectacular.

FARM EXHIBITS. Throughout Roger Williams Park Zoo's 150-year history, it has exhibited a variety of animals—local native species, species from around the world, and domestic species. Farm animal exhibits, particularly compelling destinations for younger audience members, have been popular since the earliest days of the zoo. From the 1980s exhibit shown above to the Full Circle Farm exhibit from 2006 pictured below, exhibits like these highlighted the importance of conserving all endangered breeds, including farm animals. Working with conservation partners throughout the state, the Full Circle Farm focused on getting back to small farming practices and preserving heritage breeds that offer diversity and specialization that modern factory farms do not.

AMERICAN BURYING BEETLE FIELD WORK, 2021. In the 1990s, the US Fish and Wildlife Service initiated a project to capture mating pairs of American burying beetles from Block Island off the coast of Rhode Island and attempt to rear them in captivity, with the aim of repopulating them in a former habitat. Roger Williams Park Zoo joined the project in 1994 and became the sole breeding facility for program. The zoo also contributed staff, data, and field support to the project, rearing over 5,000 beetles and releasing almost 3,000 on the island of Nantucket, Massachusetts. In 2006, the Association of Zoos and Aquariums created a Species Survival Plan (SSP) for the American burying beetle—the first SSP for a terrestrial invertebrate. Lou Perrotti (standing at left in the above image), the zoo's director of conservation programs, directed this program, which has since expanded to two other zoos and to other invertebrates.

STATE INSECT LEGISLATION SIGNING, 2015. Inspired by the efforts of Roger Williams Park Zoo conservation director Lou Perrotti, the director of the American burying beetle Species Survival Plan, a group of Rhode Island elementary school students successfully spearheaded a campaign calling for the American burying beetle to be named the official state insect. Gov. Gina Raimondo is pictured at the signing event in 2015. (Courtesy of Rhode Island Senate.)

JANE GOODALL VISIT, 2010. Roger Williams Park Zoo has several connections to famed primatologist Jane Goodall. On May 1, 2010, as part of a 50th-anniversary tour to honor her groundbreaking research with chimpanzees in the wild, Goodall visited Roger Williams Park Zoo. During her visit, she also met with members of the zoo's local chapter of Roots and Shoots, a youth conservation program that she founded in 1991. Goodall dedicated a full chapter of her 2009 book *Hope for Animals and Their World: How Endangered Species Are Being Rescued from the Brink* to the work the zoo is doing to save the American burying beetle.

VETERINARY HOSPITAL (ABOVE) AND ARMADILLO PROCEDURE (BELOW), 2011. The zoo's veterinary facility had occupied the basement level of the Sophie Danforth Center, itself a former barn, until 2011, when the John J. Palumbo Veterinary Hospital opened on zoo grounds. The new facility occupied 50 percent more square footage than the previous facility and had dedicated rooms for exams, surgery, necropsies, and X-rays, as well as a laboratory and pharmacy. Although it is located on site, it is in a secluded area toward the perimeter of the zoo with a dedicated service road and exterior gate.

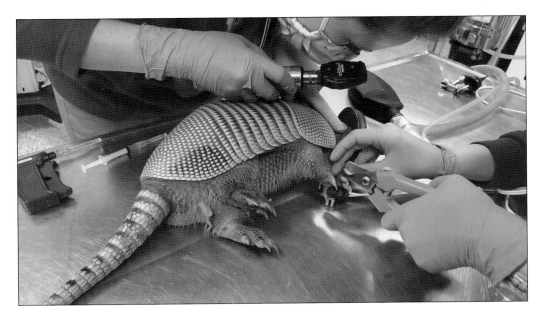

SNOW LEOPARD PROCEDURE AND ELEPHANT X-RAY. With over 160 species of animals as of 2022, the zoo's veterinary team must be prepared to provide care for a variety of animals, sizes, and situations. Diets, medications, routine examinations, and emergencies require specialized personnel and equipment. Zookeepers are the first line of care for each animal. In addition to being responsible for the animals' day-to-day care, keepers are typically the first people to observe changes in behavior—such as an appetite change or lethargy—that might indicate a medical issue. While the John J. Palumbo Veterinary Hospital is equipped to accommodate very large animals, some animals require that equipment be brought to them, as with the elephant X-ray pictured below. Since most of the animals are trained to assist in their own care (by presenting their body parts, for example), keepers and veterinarians can more easily administer medical care at the exhibits.

HASBRO'S OUR BIG BACKYARD, 2012. As a result of social science research, the zoo set out to create an exhibit that connected with kids at their own developmental level, knowing that it is the most reliable way for children to establish a solid relationship with the natural world. Hasbro's Our Big Backyard opened in 2012 and is geared toward the way kids learn naturally—with free choice and open-ended play.

CHILDREN AT HASBRO'S OUR BIG BACKYARD. The outdoor play and exploration area of Hasbro's Our Big Backyard included the CVS Health Treehouse; Our House, an education program space; a greenhouse for education programs and events; and the Nature Swap, a "trading post" that encourages participants to share items found in nature or photographs, artwork, or journal pages that reflect a personal experience with nature.

CHILD IN OUR BIG BACKYARD TREEHOUSE. Children who visit Hasbro's Our Big Backyard are guided by specially trained Play Partner volunteers who help them make the most of their visit. Offering a variety of exploratory play experiences, including loose items, water play, and mud play, Our Big Backyard is often the first association that younger kids have with the zoo, making it critical for fostering love and respect for the natural world.

WATER PLAY AT OUR BIG BACKYARD. Because it is owned by the City of Providence and operated by the Rhode Island Zoological Society, the zoo's connection with Providence is an important one. Not only is the zoo Rhode Island's most popular tourist attraction, it is also the primary connection to nature for many city residents. As a thank you to the people of Providence, the zoo offers free admission to residents on the first Saturday of each month.

BEYOND THE FENCE NATURESCAPE. In 2013, Hasbro's Our Big Backyard expanded into the Beyond the Fence Naturescape area. This area is more spread out than Our Big Backyard and gives children the opportunity to use more than just their hands. Climbing, swinging, digging, fort-building, and other fully active play is encouraged in this part of the zoo.

WATER WALL AT OUR BIG BACKYARD. Although Rhode Island is cold in the winter, it also gets hot in the summer. Our Big Backyard is home to the Water Wall and water play areas, which became a big hit with children who want to cool off and parents who want a break from walking. All are eager to soak just their fingers—or perhaps themselves from head to toe.

SANTA'S ARRIVAL AND PARADE, 2013. As a year-round zoo, Roger Williams Park Zoo has had occasion to host many holiday celebrities. Santa Claus has paid visits to the zoo during holiday parades, and both he and the Easter Bunny have made appearances for photo opportunity meet-and-greets. When the zoo acquired the operating rights to the park's carousel in 2016, many of the zoo's holiday functions were shifted to that location.

ALEX AND ANI FARMYARD, 2014. The Alex and Ani Farmyard, opened in 2014 and designed to resemble a typical Rhode Island farm, celebrates the diversity of farms around the state. With over 1,200 farms, agriculture is a particularly important industry in Rhode Island. Younger visitors, especially those born and raised in cities like Providence, might not be aware that the eggs they eat and the milk they drink come from animals, not containers. The graphics and play opportunities within the exhibit immerse families in farm life, teaching them about a farmer's day-to-day work, the reasons they love farming, and the challenges they face as farmers.

FEEDING GOATS IN WINTER. With the focus of the Alex and Ani Farmyard being on local farming, the zoo collaborates with partners from around the state. Children partake in play-based activities such as a milking station sponsored by Monroe Dairy and an egg-collection station sponsored by Little Rhody Farms. Tall, child-sized graphics depict children speaking with local farmers, and the accompanying text paints a picture of the farm lifestyle.

CHILDREN IN THE CONTACT YARD. After a day of observing a variety of wild species, the appropriately named Contact Yard within the Alex and Ani Farmyard allows kids (and adults) to interact with animals on a more personal level. While sharing a space with goats (and sheep, during the yard's first few years), visitors can pet and brush the animals. Food dispensers are also provided outside the exhibit and accounted for in the animals' diets, so children who are unsure about entering the space with the animals can still have a close encounter with them. Alpacas, rabbits, chickens, and a donkey round out the farm's collection.

ALEX AND ANI FARMYARD ENTRANCE, 2014. The Alex and Ani Farmyard is a prime example of the type of strategies Roger Williams Park Zoo hoped to implement when developing new exhibits and programs for its 20-year master plan that began in 2015. Each new area of the zoo would need to incorporate multiple components to make it more than simply a themed area in which to exhibit animals. Most significantly, there would be an associated "ask" or call to action for each area. In the case of the Alex and Ani Farmyard, visitors are encouraged to buy local produce. In the Faces of the Rainforest exhibit, which opened in 2018 as the first exhibit in the master plan, visitors were asked to examine their buying habits and daily routines with an eye toward how they affect— and are affected by—the rainforests.

ALEX AND ANI FARMYARD AND OPENING TEAM, 2014. Members from different departments throughout the zoo contributed, time, resources, and labor to help bring the largely in-house project to completion. Pictured are, from left to right, executive director Jeremy Goodman, director of education Shareen Knowlton, deputy director of operations Ron Patalano, David Albaugh, Brett Cortesi, James Festa, Lynne McClain, Christopher Lamoureux, Jan Drugge, Leigh Picard, Sean Brouillette, Benjamin Feliciano, Justin Blackwood, Peggy Ogert, Jesse Carter, temporary laborer David ?, Ian Didow, Joseph Brightman, Nicholas Branch, and Jayson Ledoux.

OUTBACK TRAIL, 2014. While kangaroos and wallabies had previously been in the zoo's collection, in 2014, the zoo developed a brand-new way to showcase them. The Outback Trail was located across from the Alex and Ani Farmyard and adjacent to Tropical America. Unlike a traditional exhibit, in which visitors observe animals through a barrier, the Outback Trail provided a unique walk-through experience for visitors, allowing them to share the same space as the animals. A simple rope fence lined the perimeter of the trail, which kept visitors from straying into the animals' area but allowed the animals to cross the path freely. The Outback Trail was closed in 2017 to allow for construction on the Faces of the Rainforest exhibit.

JUNIOR DOCENTS, 2016. The zoo has supported a wide variety of volunteer programs through the years, many of them youth-focused. These have included Play Partners, junior keepers, ZooCamp counselors in training, and junior docents. The zoo has seen many individuals grow up connected to it in one way or another throughout their lives. It is not uncommon for children to begin an association with the zoo starting as a four- or five-year-old ZooCamper attending Tadpole Academy, then experiencing each level of ZooCamp as they grow older, eventually becoming counselors in training and, in some cases, full-time employees as they approach adulthood. Pictured here are junior docents Michael Gibson-Prugh (left) and Shania Hardy.

GREENHOUSE WITH EVENT SETUP. The greenhouse, located adjacent to Hasbro's Our Big Backyard, was designed as a flexible space. It has hosted seasonal animal exhibits such as the Flutterby butterfly garden (2014); Birds from Down Under budgie-feeding experience (2016); and Shades of Nature, featuring a white alligator (2019). When not in use as an animal exhibit, it has served as one of the zoo's reservable event spaces.

PRIVATE EVENT AT THE GREENHOUSE. Over the years, Roger Williams Park Zoo has expanded its appeal to visitors by serving as a spectacular venue for hosting anniversaries, birthday celebrations, showers, or other special events. A variety of indoor and outdoor locations are available for both daytime and nighttime events. Pictured is an event scheduled in conjunction with the Jack-O-Lantern Spectacular.

BIRDS FROM DOWN UNDER, 2016. With the closure of the Tropical America exhibit in preparation for the creation of the Faces of the Rainforest exhibit (slated to open in 2018), it was important that the zoo was able to offer a new experience for visitors in 2016. Since exhibits typically take years to design and construct, developing a full-scale replacement exhibit on a relatively tight timeline was not a viable option. However, the zoo was able to contract with a third-party provider, Living Exhibits, that supplied budgies and parakeets as well as the training for staff to facilitate this interactive bird-feeding experience.

CAROUSEL ATTRACTIONS, 2016. Although the carousel had existed as part of the park since 1897, when the zoo consolidated its collection in the 1960s, it resulted in the carousel being located about a quarter mile away from the zoo proper. The carousel was managed by a third party until 2016, at which point the zoo acquired the operating rights. With the zoo now managing the carousel, it was able to shift many events to the new venue, freeing up valuable zoo resources. In addition, it is able to host new events such as the extremely popular Food Truck Friday (pictured above).

WOODLANDS EXPLORER TRAIN. The Woodlands Explorer was purchased by the zoo in 2017 and integrated into the Explore & Soar area. The train, which is popular with the youngest visitors and parents looking for a break from the busier areas of the zoo, follows a trail through a peaceful wooded grove adjacent to the Wetlands Trail, an area that allows visitors to view native animals in their natural habitat.

CAMEL RIDE, 2018. Adjacent to the Soarin' Eagle zip line ride and Wilderness Explorer train ride, the zoo offered camel rides for a few seasons beginning in 2014 with the intent of providing visitors with an opportunity to experience animals differently than they would at an exhibit. The camels, owned and operated by an outside company, were not part of the zoo's animal collection.

Soarin' Eagle Ride. In 2017, the zoo opened the Soarin' Eagle zip line ride, the first modern thrill attraction at the zoo. Located in the Explore & Soar area, this exciting ride gives visitors a view of the zoo from 115 feet above Polo Lake. Riders who shift their gaze higher are also treated to a view of the downtown Providence skyline.

Brew at the Zoo, 2018. The annual adults-only after-hours Brew at the Zoo has been an eagerly anticipated sellout event since 2014. Attendees can sample more than 170 craft beers, ciders, and other beverages from more than 80 local, regional, and national brewers and food from festive food carts and cafés, dance to local bands, and enjoy a variety of close-up animal encounters.

FACES OF THE RAINFOREST OPENING, 2018. The zoo's most recent major exhibit is Faces of the Rainforest. Planning for the project began in 2015, and it opened in November 2018. Unlike many projects of the past, it was designed with its interpretive messaging in mind from the beginning. Historically, educational content was developed after an exhibit had already been planned and designed. In this case, the educational content helped inform the building design and selection of inhabitants. Because the zoo wanted an exhibit that would highlight every individual's connection to the rainforest, regardless of how far away they are from it, the zoo conducted visitor interviews to gauge people's familiarity with the rainforest and its challenges. Those findings helped the zoo focus the exhibit's educational messaging.

"You cannot get through a single day without having an impact on the world around you.

What you do makes a difference, and you have to decide what kind of difference you want to make."

- Jane Goodall

JANE GOODALL QUOTE. One of the primary goals of Roger Williams Park Zoo's Faces of the Rainforest exhibit, which opened in 2018, was to convey to visitors how a place as seemingly far away as the rainforest can still have a direct connection to the everyday lives of people in Rhode Island. This quote from Jane Goodall conveys that concept perfectly, so the zoo asked Goodall for permission to use it at the exhibit. Permission was kindly granted, and a dynamic five-by-seven-foot sign was constructed, resulting in a dramatic welcome to the exhibit that hundreds of visitors have their picture taken with each year. The quote on the sign says: "You cannot get through a single day without having an impact on the world around you. What you do makes a difference, and you have to decide what kind of difference you want to make."

FACES OF THE RAINFOREST INTERIOR. The Faces of the Rainforest exhibit's massive glass front not only provides enough sunlight for the numerous living plant species within but helps create a natural, vast environment that simulates the outdoors. Free-roaming sloths, tamarins, and colorful birds share the same space as visitors, while giant river otters, tamanduas, and a variety of monkeys are in separate habitats within the building. A splashing waterfall and birdsong complete the immersive environment. Meanwhile, through a series of graphics along the walking path, visitors learn about the variety of people who rely on the rainforest and are often surprised at the similarities they find despite the geographical distance between them.

POLLINATOR GARDEN, 2019. In 2019, Roger Williams Park Zoo hosted a "Pollinate New England" lecture conducted by the New England Wildflower Society, an organization dedicated to raising awareness of the decline of pollinating insects, birds, and other animals. The zoo was one of two sites that received a grant to install its own pollinator garden. The work was done by a team of zoo staff and volunteers.

YOGA WITH THE ELEPHANTS, 2019. As a conservation-based organization, the zoo has recognized the importance of fostering and maintaining connections with members of the community. A sense of belonging is critical to developing long-term relationships with groups and individuals. Offering a variety of experiences to different people widens the zoo's appeal and creates opportunities to increase support for the zoo. Yoga with the Elephants, Breakfast with the Animals, wine and paint nights, Insider Tours, private animal encounters, and other specially designed experiences broaden the zoo's audience and grow its support base.

Six

2020
The Pandemic

When the COVID-19 pandemic hit in early 2020, the zoo had to quickly shift gears. It was obvious that as a venue that hosts thousands of people—often in close quarters—every day, the zoo had to do something different. How could it continue to be a place that depends on people for its support and remain a enjoyable destination? It was no surprise that the zoo was able to be nimble, make rapid changes, and accommodate this new normal.

With ZooCamp and classroom lessons already scheduled, the education department had to quickly come up with an alternative plan for these extremely popular activities—and it did. For the first time, the zoo incorporated distance learning. It received a grant from the Institute of Museum and Library Services to explore and develop ways to support classroom teachers, as their needs were changing as well. Even though the zoo prides itself on tangible experiences, these virtual lessons often had perks that an in-person visit could not offer, such as seeing an elephant's mouth up close through a keeper's phone camera. As a result of these virtual experiences, Roger Williams Park Zoo was no longer limited to local programs, and it became an online destination for schools all over the country.

The zoo also implemented social distancing signage, one-way routes and maps, and more outdoor activities. Whether it involved turning outdoor events into drive-through experiences or setting a limit on the number of people allowed inside an indoor space at one time, the zoo was able to not only continue but to thrive.

Sentinel Dog Statue, Winter 2021. Like many businesses, when the COVID-19 pandemic hit in 2020, the zoo needed to find ways to adapt. During a short closure for the first few months during the zoo's busiest time of year, the zoo developed safe ways to continue to manage the animal collection, maintain the grounds, provide educational programs, and hold annual events. The pandemic plan developed by the zoo was heralded as exemplary and was subsequently adopted by a variety of institutions.

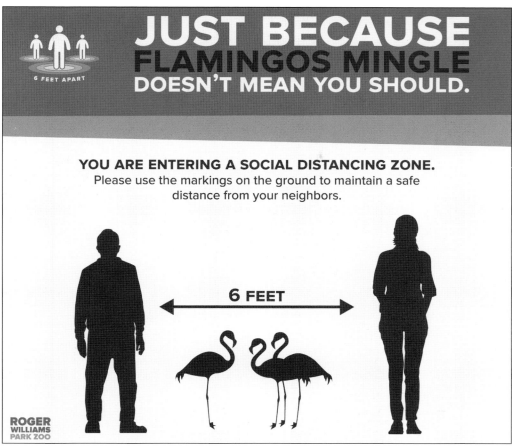

FLAMINGO SOCIAL DISTANCING SIGN, 2020. The zoo reopened in June 2020, adhering to COVID-19 safety protocols instituted by the Centers for Disease Control and the Rhode Island Department of Health. As such, the zoo had to develop an entirely new system of daily operation that would ensure the safety of guests and animals. Several indoor exhibits, including Faces of the Rainforest and World of Adaptations, remained closed.

AMERICAN BURYING BEETLE SOCIAL DISTANCING SIGN, 2020. The use of outdoor areas that would normally involve groups of people in a limited space, such as Our Big Backyard, Wild Bunch performances, and the farm's Contact Yard, was also suspended. A timed entry system was developed to ensure that crowd sizes could be managed to comply with national and local COVID regulations.

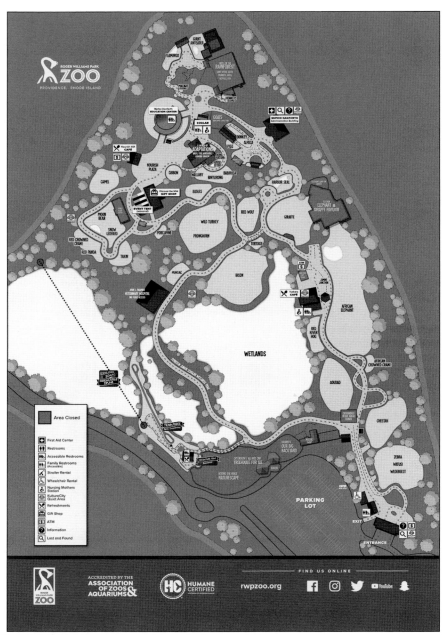

One-Way COVID Map, 2020. Although the timed entry system was meant to ensure that the zoo's capacity would not exceed safe limits, crowds still needed to be managed in such a way that visitors could comfortably move through the zoo while minimizing exposure to other visitors, even in outdoor spaces. The zoo's unique "upper and lower loop" layout worked to the its advantage. A counterclockwise flow was implemented with an optional bypass where the two loops met, allowing guests to shorten their visits or repeat entire sections of the zoo without having to travel "upstream." With shows and other crowd-based experiences suspended, the interpretive staff who would normally facilitate those programs, and who were already highly skilled in guest interactions, were stationed throughout the zoo to assist visitors and enhance their experiences despite the unusual circumstances.

Virtual Educational Programs, 2020. In addition to the challenges the coronavirus created for day-to-day operations, animal care, and staffing, significant revenue generators such as events, camps, and other programs could no longer operate as they previously had. The zoo quickly developed alternate educational programming for an audience that still craved a connection to the zoo—perhaps even more so due to the effects of a prolonged lockdown. As a nonprofit organization, the Rhode Island Zoological Society knows that every day the zoo's gates are closed or a program cannot run can have a profound impact on the ability to care for animals and keep staff employed. Roger Williams Park Zoo was one of the first zoos to develop a virtual camp program and eventually hosted webinars to advise other zoos.

DRIVE-THROUGH JACK-O-LANTERN SPECTACULAR, 2020. When the zoo reopened in June 2020 with COVID-19 protocols in place, the zoo management knew that the annual Jack-O-Lantern Spectacular, immensely popular and the zoo's largest event-based revenue generator, might be in jeopardy. Normally a leisurely walk-through event, each night, it attracted thousands of people to stroll along the winding path to admire the carvings. Planning for the October event typically begins months in advance, so because the future was so uncertain, the event's organizers developed a unique solution: make the spectacular a drive-through event during which families can remain safely in their vehicles and still enjoy the show.

DRIVE-THROUGH JACK-O-LANTERN SPECTACULAR FINALE, 2020. To accommodate vehicles during the new drive-through Jack-O-Lantern Spectacular, event organizers shifted the displays from their usual homes on the Wetlands Trail to the main path that runs from the zoo's main entrance through the African exhibit area and along a service road closed to guests during normal operation. This was the first year that the pumpkin displays were located on paths accessible by guests during the day, making daily maintenance of the delicate displays even more challenging. Another first for 2020 was the addition of Skeleton Lane, a grand finale after the main pumpkin display featuring a variety of skeletons celebrating Halloween in their own festive way.

DRIVE-THROUGH HOLIDAY LIGHTS SPECTACULAR, 2020. As businesses began to reopen during the COVID-19 pandemic, Rhode Island looked for ways to bring tourism dollars back into the state. Hotels and restaurants needed to give people a reason to come to Providence. After successfully developing and delivering a popular drive-through Jack-O-Lantern Spectacular, the zoo decided to do something similar for the winter holidays. The light displays and supporting infrastructure were paid for by a grant from Rhode Island Commerce's Hotel, Arts and Tourism (HArT) Recovery Grant Program, and the zoo would be able to retain both for future shows. The Holiday Lights Spectacular became a new annual zoo event, and with COVID restrictions beginning to ease, the second iteration became a walk-through event in 2021.

Seven

2021 AND BEYOND
THE ZOO OF THE FUTURE

Looking to the future, the zoo plans to continue its focus on inclusion, state-of-the-art husbandry, community involvement, and creating a culture of conservation within the organization. On the horizon is a new Education Center that will hold a licensed preschool, giving the area's youngest community members a chance to become zoo kids. The building itself is being designed to align with the zoo's conservation mission in that it will be an energy-efficient, net-zero-energy building. The zoo looks to build upon its successful animal care practices with even more rigorous assessment and evaluation. The importance of inclusion continues to have an impact on the zoo's exhibit design and educational programs, as well as internship programs, centering the knowledge that people from all walks of life can become environmental stewards.

The Green Team, which consists of zoo staff members from various disciplines, leads internal and community initiatives that empower individuals to make a difference through projects that range from recycling cellphones to save gorillas' habitats to composting food from zoo events. As the zoo evolves, each area is slated to include a call to action to let the public know that they can make a difference, that everyone is part of the mission, and that zoos alone cannot save the world.

WHITE ALLIGATOR, 2021. The greenhouse, home to temporary seasonal exhibits such as Flutterby (2014) and Birds from Down Under (2016), became home to Shades of Nature in 2021. This exhibit highlights unusual animal colorings in nature and features Elsa, an albino alligator. Originally slated for a 2020 arrival but delayed due to the pandemic, the exhibit has proved to be popular with visitors, many of whom were visiting the zoo for the first time since the COVID-19 pandemic lockdown began.

ASIAN LANTERN SPECTACULAR, 2021. Although the Jack-O-Lantern Spectacular and the Holiday Lights Spectacular shows were unqualified successes, they both took place towards the end of each calendar year, which meant the spring was wide open for a similar event. In early 2021, on the heels of the first drive-through Jack-O-Lantern Spectacular and the first Holiday Lights Spectacular, the zoo developed its new springtime show, the Asian Lantern Spectacular. The zoo contracted with Hanart Culture, producer of Asian lantern shows throughout the world. Unlike the Jack-O-Lantern and Holiday Lights Spectaculars, each of which ran for an entire month, the Asian Lantern Spectacular ran from mid-April to July 4—a significantly longer run. The show returned in 2022 for another installment featuring brand-new displays mixed with returning favorites.

ZOOKEEPERS, 2019. This picture includes a small sample of the zoo-keeping team that ushered Roger Williams Park Zoo through the pandemic of 2020 and into the future. While the full team consists of over 30 members, each with their own area of expertise, they are represented here by, from left to right, (first row) Becca Hollenbeck, Steve Skitek, and Laura Isaacs; (second row) Kim Warren and Anne Tan; (third row) Melissa Ciccariello, Kelly Froio, Thomas Troy, Anthony Polite, Gabriel Montague, Vic Froio, Brett Haskins, Lisa Ruggiero, and Christine MacDonald.

YOUTH ADVISORY COUNCIL, 2019. In 2019, the zoo formed the Youth Advisory Council. Consisting of a small but dedicated group of high school and college volunteers, the council was created with the goal of providing a youth perspective on existing and planned zoo programs. The zoo's directors and managers could pitch ideas or questions to the group and solicit feedback. In addition, the group could participate in the zoo's offerings as guests would and report back their findings and offer suggestions for improvement. With an eye toward helping the zoo attract and engage a young-adult audience, the council also offered input on creating robust and engaging events for youths. Pictured here are, from left to right, Emerson Berriman, Hannah Wood, Katie Kuhl, Corey Pierce, Julia Figueroa, Jaden McCarthy, and Olivia Massotti.

SENTINEL DOG STATUE FAMILY PHOTOGRAPHS. Having photographs taken on or near the *Sentinel* dog statue is a long-standing tradition for families, staff, and volunteers. Multiple generations of visitors have documented their trips to the zoo through the years by ensuring they take new photographs during each visit despite the challenges of comfortably sitting on a bronze statue in cold winters or hot summers. The statue was initially located in Roger Williams Park and eventually moved to within the zoo's boundaries as the animal collection was consolidated into a more centralized area. On April 1, 2022, a time capsule was buried beneath the statue as part of the zoo's 150th birthday celebration.

ROGER WILLIAMS PARK ZOO ENTRANCE. What began in 1872 as a collection of small animal exhibits throughout a newly formed Roger Williams Park is now a worldwide destination and award-winning zoo accredited by the Association of Zoos and Aquariums and the American Humane Society. In 2021, Roger Williams Park Zoo welcomed approximately 834,000 visitors to its 40 acres and featured over 100 different species of animals.

INDEX

About Roger Williams Park Zoo

Roger Williams Park Zoo is part of the heart and soul of Rhode Island along with frozen lemonade, hot wieners, sandy shorelines, and the Blizzard of '78. Visits to the zoo are part of growing up in Little Rhody. Most native Rhode Islanders have a photograph of someone in their family sitting on the *Sentinel* dog statue—a memory that will often stay with people for their whole lives. Numerous kids that started in the zoo's camp or preschool programs are now professionals at various zoos and aquariums, leading others in the world of conservation.

As more than just a wonderful family destination, this little zoo in the nation's smallest state has transformed over time. What began as a place where people could get a close look at a small collection of animals is now a world-class conservation, education, and recreation organization. Although it is known as a smaller zoo that does not think it is small, Roger Williams Park Zoo is a globally recognized major player with a strong reputation.

Because of Sophie Danforth's compassion, vision, and tenacity, the zoo became more focused on animal welfare, hired its first professional director, and developed educational programs starting in the 1960s. So deep was her passion that she was hands-on in doing whatever it took to keep the zoo going and growing, whether it was donating funds, creating and running the gift shop, negotiating with unions, working a small printing press to print her own signs, or taking animals into classrooms. Danforth's strong belief in conservation was clear when she said that she "understood the importance of each animal in this world and its niche and how important it is that each niche be preserved."

Roger Williams Park Zoo, operated by the Rhode Island Zoological Society, carries on Sophie Danforth's conviction with a strong focus on animal welfare and care. The society is also deeply dedicated to engaging audiences and empowering them to act on behalf of wildlife while having fun. The nonprofit organization understands the central role the community plays in conservation and feels passionate that every individual can make a difference. The zoo's mission aims to "empower guests to join us in conserving wildlife and wild places" as it works to improve the world while creating a destination for great times and wonderful memories.

DISCOVER THOUSANDS OF LOCAL HISTORY BOOKS
FEATURING MILLIONS OF VINTAGE IMAGES

Arcadia Publishing, the leading local history publisher in the United States, is committed to making history accessible and meaningful through publishing books that celebrate and preserve the heritage of America's people and places.

Find more books like this at
www.arcadiapublishing.com

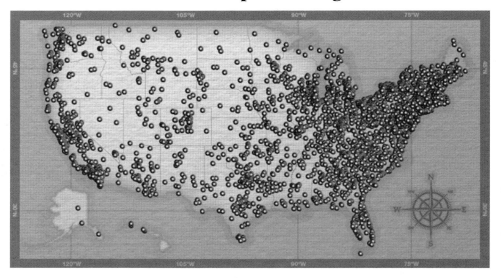

Search for your hometown history, your old stomping grounds, and even your favorite sports team.